YOUR KNOWLEDGE HAS VALUE

- We will publish your bachelor's and master's thesis, essays and papers

- Your own eBook and book - sold worldwide in all relevant shops

- Earn money with each sale

Upload your text at www.GRIN.com and publish for free

Alternate Theory for the Big Bang and Standard Model. How are Galaxies created?

Daniel Stark

Bibliographic information published by the German National Library:

The German National Library lists this publication in the National Bibliography; detailed bibliographic data are available on the Internet at http://dnb.dnb.de.

ISBN: 9783346778666
This book is also available as an ebook.

© GRIN Publishing GmbH
Nymphenburger Straße 86
80636 München

Print and binding: Books on Demand GmbH, Norderstedt, Germany
Printed on acid-free paper from responsible sources.

The present work has been carefully prepared. Nevertheless, authors and publishers do not incur liability for the correctness of information, notes, links and advice as well as any printing errors.

GRIN web shop: https://www.grin.com/document/1306358

Daniel Lee Stark
August 5, 2022

Alternate Theory for the Big Bang and Standard Model

Abstract

Alternate theories for the Big Bang and Standard Model are proposed which theorize how galaxies are created. The galaxy formation model explains how and where the fundamental particles are created.

The greater universe is theorized as consisting exclusively of Dark Matter (DM) and Energy. DM has mass and is the source of the gravitational field. DM hosts kinetic and potential energy. A stable greater universe is the end result of entropy wherein DM and energy are uniformly distributed.

Fluctuations in these evenly spaced DM particles allow gravitational attraction to concentrate DM creating galactic size DM strings. The masses of DM violate entropy by concentrating kinetic and potential energy. In these DM concentrations, collisions between large DM concentrations form a spinning DM galaxy. Inside these spinning turbulent DM galaxies, DM is theorized to undergo phase change into particle size droplets. DM droplet sphere collisions are theorized to create gamma rays by converting kinetic energy into gamma ray photons (probably via 3rd or 4th derivative of DM sphere velocity).

DM spheres are predicted to capture gamma rays via Total Internal Reflection (TIR) creating a DM sphere with EM fields. Thus creating electrons, positrons, neutrons and protons. TIR is possible when the index of refraction between inside the DM sphere and outside satisfy: $(n_1/n_2)(\sin \theta_1) > 1$ where $n2 = 1$ which is outside of the DM sphere, n_1 is the DM's index of refraction and θ_1 is the angle from vertical to the DM's surface. These gamma rays may be generated inside of the DM sphere during the collisions or be captured because of the DM sphere's expected high index of refraction.

To create a particle with EM forces, the particle must support internal standing waves which requires the DM sphere dimension to satisfy gamma ½ wavelength dimensions. The simplest being a half wavelength diameter sphere which probably results in an electron or positron when capturing a gamma ray.

The TIR photon provides a total positive on negative external $\mathbf{E}$ field because the trailing photon's $\mathbf{E}$ field is reversed by the lens effect during reflections inside the DM sphere yielding

i

an **E** field in the same direction as the first half of the photon's **E** field direction.

Conclusions:

DM and energy are theorized as the fundamental building blocks of all matter and are expected to be independently conserved.

The Big Bang never occurred. Each galaxy internally creates the fundamental particles from DM and energy which are electrons, protons, etc., allowing for a galaxy formation.

Charge is created as explained by Dr. Milo Wolff, by outgoing or incoming **E** fields.

Charge is responsible for particle symmetry, e.g. an electron or positron. Super symmetry is redefined as simple symmetry.

DM replaces our present concept of virtual particles.

Table Of Contents

Table of Figures

1.1. Introduction

This paper first describes what is known about energy and DM.

This theory introduces a new photon concept, based on photon holograms and images.

The described photon model supports Dr. Milo Wolff's theory that ingoing and outgoing electric fields create the negative and positive electrical charge; however, he lacked a credible model. This paper provides a potential Wolff model which morphs into an alternate for the Standard Model.

1.2. DM and Energy Theory

Energy is theorized to only exist associated with DM. DM has mass and provides a gravitational field. Thus energy can be in the form of kinetic or potential energy directly associated with DM. Photons are also associated with DM in a more complex manner extensively discussed later.

1.3. DM Index of Refraction

DM is observed to possess an index of refraction evidenced on the gamma ray time of arrival waveband spread from deep space gamma ray bursts.[1]

The mechanism for a DM index of refraction cannot be explained by current theories because current index of refraction requires electrons. DM does not have electrons.

1.4. Photons

Photons can also carry energy associated with DM. Photons energy should be calculated by the sum of an oscillation energy plus kinetic energy. Currently, the total photon energy is simply calculated as $E=hc/\lambda$.

Abrupt changes in electron momentum, $\mathbf{p}$, are observed to create photons.

Photons are observed possessing kinetic energy and momentum, presumably having to conserve the momentum and kinetic energy during the electron's abrupt velocity change.

1.4.1. Gamma Ray Sources

The center of most galaxies contain a large concentration of DM including our own galaxy. The center of our DM galaxy is also a strong gamma ray source.[1] These gamma rays are theorized to be produced via DM "droplet" collisions wherein the DM gas phase changes into droplets of DM resembling a solid or liquid sphere.

Photons are created by electrons when subject to kinetic energy changes. The kinetic energy and momentum are hosted by a photon, which explains why a photon must have a velocity.

If electrons are also composed of DM as theorized herein, DM in the electron creates the gamma rays. A detailed photon description follows later in this paper.

1.5. DM Phase Change

All matter we currently observe can have several phases, gas, liquid or solid. It is reasonable to suspect DM is not unique; therefore, DM is theorized to have a gas, liquid or solid phase. DM is also theorized to be of particulate nature and attractive but obeying the Pauli Exclusion principle wherein two DM particles cannot occupy the same point in space. The conclusion is based on lack of singularities. If DMs could occupy the same space, they could form singularities, which are not observed.

The interior of a large DM concentrations such as in the center of galaxies, offers high density DM conditions.

Inside these DM concentrations, DM is theorized to undergo a phase change from a gas to particle size liquid or solid sphere. The spheres are of various sizes analogous to water vapor forming spheres. These DM droplets are considered spherical and transparent with a high index of refraction. The density, not optical density, can be estimated by calculating the electron mass and quarter wavelength[1] of a 10^{-12} meter gamma ray to determine the spherical volume yielding 13,918,113 kg/m^3. This value can be compared to uranium's density of 19,050 kg/m^3, which is 730 times more dense than uranium indicating a very high index of refraction since density and optical density are associated.

Various sizes of DM spheres are expected to form up to some environmental limit.

DM sphere collisions in the center of the DM concentrations are theorized to convert kinetic energy into gamma rays in the same manner as electrons. The center of our galaxy is a know gamma ray source.[1]

1.6. Photon Discussion

This thesis is also based on a different conceptual photon description.

The photon's energy oscillates between the **E** and **M** fields. The visual photon's concept is an oscillator wherein the **E** fields oscillates between outgoing and ingoing **E** fields with the **M** field oscillating in the same manner out of phase with the **E** field and at 90° angled position. Typical photon drawings show a sine wave with the ingoing and outgoing **E** field

---

[1] *The 10^{-12} meter wavelength is based on electron positron decay to gamma rays.*

---

separated between the first and second part of the photon wave along the poynting vector, but only in one direction e.g +x direction. The new concept describes the **E** field forming a photon to be in the +x and -x directions simultaneously. A saw tooth coincident with an inverted saw tooth **E** waveform best describes the concept shown in Figure 1. The **M** waveform, not shown, is 90° out of phase and perpendicular to the **E** waveform. Figure 1 shows the **E** vector(neglects **M** vector) for both concepts.

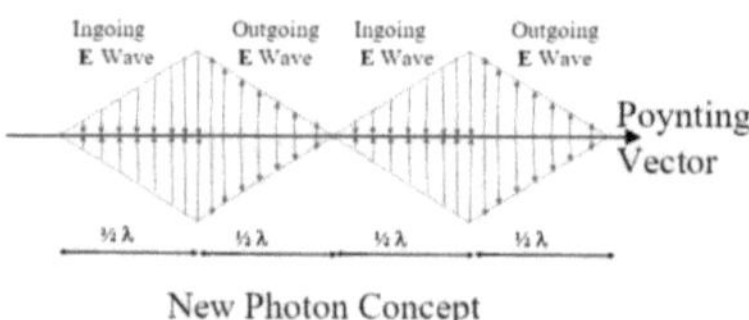

New Photon Concept

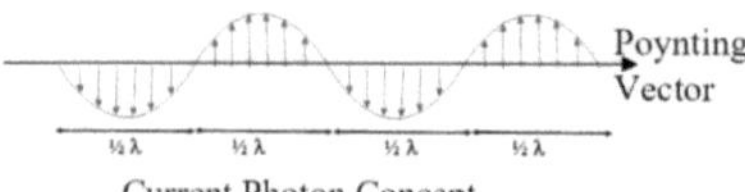

Current Photon Concept

Figure 1 New & Current Photon Concept

In reality only part of the wave exists at any instant of time. Thus the photon is a total outgoing **E** field with a decaying **M** field followed by a totally ingoing **E** field and an increasing **M** field. The typical sine wave alternates the direction of the **E** wave sequentially rather than considering a spherical wave that

alternates between ingoing and outgoing **E** and **M** fields from a central position. Both concepts depict a photon over a period of time, not an instant in time.

The photon has dimension thus it is not a point; however, it is possible to describe the conditions at a point mathematically, and the photon has a center. The photon is also three dimensional with the electric field, **E**, perpendicular to the magnetic field, **M**. These fields are oscillating and extend outward from the center of travel - Poynting vector. The photon acts as an oscillating particle with extended **EM** fields.

Dr Milo Wolff states a sine wave does not correctly describe the photon. Dr Milo Wolff suggests a log wave shape-of *spherical waves*.

The center of the photon can go through a single slit, but the fields associated with that photon extend through the adjacent slit, causing diffraction. The photon may not be a particle, but an energy packet with dimension. How a photon contains kinetic energy is an issue.

Photon's should not be compared to waves on a water surface. Waves on the surface of water describe a wave at the interface between water and air. A more correct picture of a water wave is a three dimensional pulse under the water. Water waves propagate via kinetic energy exchange between adjacent water molecules and transfer momentum. Photon propagation is similar, but is directed in a particle fashion unlike the spherical water wave.

1.6.1. Photon Hologram & Image

A photograph of a photon using a new hologram technique (two-photon quantum interference)was collected by University of Warsaw, Dr. Radoslaw Chrapkiewicz[2]. Figure 2 is based on the photograph showing photon wave patterns. The orange/yellow arms are the oscillating electric fields. The blue field is the oscillating magnetic field 90^0 out of phase with the **E** field. The photon is a complex oscillator with a central standing wave-center and four oscillating traveling wave arms with two being the electric field and two perpendicular to the electric field being the magnetic field. The photograph is available at website:
https://cosmosmagazine.com/science/physics/what-shape-are-photons-quantum-holography-sheds-light/

Chief Physicist Fabrizio Carbone at the École Polytechnique Fédérale de Lausanne in 2015 used a lasers illuminating a nanowire in opposite directions to create a photon standing wave which excited electrons. The excited electrons in the standing wave radiate providing an image of the photon standing wave. An article describing the experiment can be accessed at:
https://www.zmescience.com/science/what-is-photon-definition-04322/

Figure 2 is an artist depiction of the results showing the **E** and **M** vector fields over one wavelength in time. The result is not a simple sine wave but a complex oscillator. The first half of the photon's pulse shows a collapsing **M** field and an increasing **E** field. The photon's **E** field is shown expanding outward up and down while the perpendicular **M** field collapses. In the second part of the photon's wave shown in Fig 2, only half of the **E** and **M** field is shown for simplicity. The **E** field is collapsing and the **M** field is increasing.

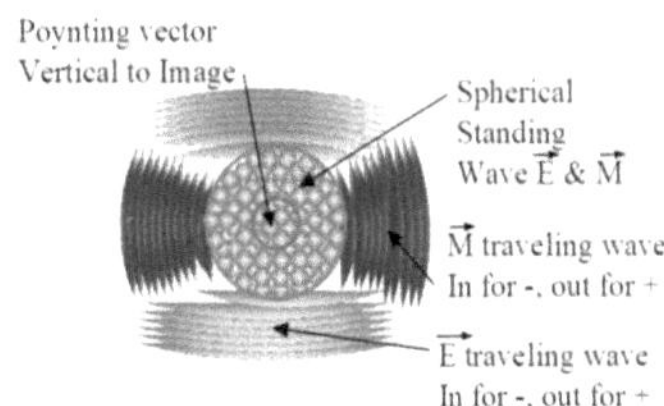

Figure 2. Photon Cross View

1.6.2. New Photon Concept

This concept differs from the current sine wave concept by showing the **E** field expanding outward in two directions with the **M** field out of phase with the **E** field doing the same. Emphasis is placed on what the photon looks like in time. At an instant in time the photon has only an outgoing **E** pulse or an ingoing **E** pulse.

An outgoing **E** vector is induced by a collapsing **M** field followed by a collapsing **E** field inducing an outgoing **M** field. The sine wave incorrectly spatially shows only half of the **E** fields during collapse and increase.

The photon is best described as an oscillator with an outgoing shaped **E** field in two directions followed by an in-going **E** field also in two directions with the oscillation energy shared with the **M** field which is angled 90° from the **E** field and 90° out of phase with the **E** vector along the poynting vector. The photon oscillator has kinetic energy and momentum requiring the photon oscillator to travel to conserve these attributes.

A photon's wavelength is composed of the sum of an increasing **E** field and a decreasing **E** field.
The second half wavelength shown in Fig. 3 shows the traced waveform over a period of time.

A photon snapshot in time is shown in Figure 4. The left side shows the photon as recorded in the hologram, the second and third photon snapshots show an oscillating pho-

ton with first an increasing **E** field and a collapsing **M** field, followed by a collapsing **E** field and an increasing **M** field. Thus the photon's pulse exists as an outgoing or ingoing pulse as shown in Figure 4, traveling spatially with time as shown in Figure 3.

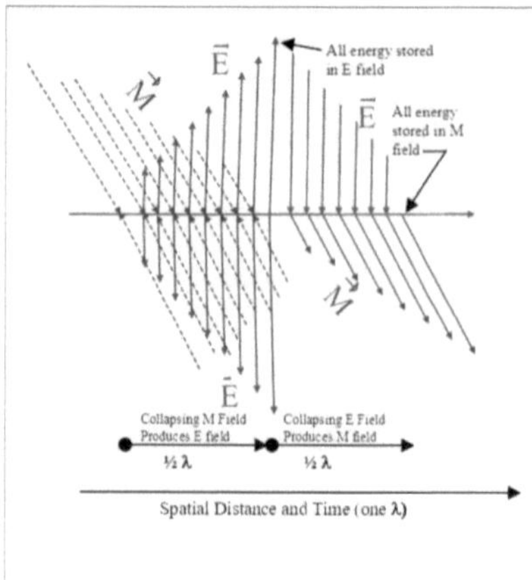

Figure 3. Photon field Pattern

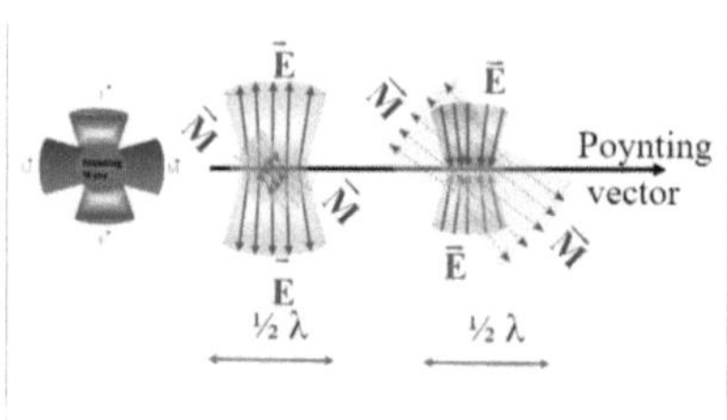

Figure 4. Photon Snapshot in Time

1.6.3. Photon Momentum

Photons do not stand still because they possess kinetic energy and momentum. Photon momentum is given by:

$$p=h/\lambda$$

Relativistic photon momentum is given by:

$$p=Ec$$

h = Planck's constant, λ=photon's wavelength, c=speed of light, E=energy[3].

The photon's kinetic energy is mathematically described as formula $K_E =h\lambda$ Where, K_E= Kinetic photon energy.

The photon's total energy is currently described as: $E=hf$ Where, E= photon total energy, f=electromagnetic frequency in Hertz(Hz).

1.6.4. Does a Photon Contain DM?

The photon's energy should be described as the sum of its kinetic energy and its oscillation energy. Or Total $E_T=h\lambda+ hf$. The open question is if a photon contains DM it would satisfy photon kinetic energy calculations and replace the rest mass theory.

The hypothesized photon DM for a 10^{-12} meter gamma can be calculated by setting: $K_E =h\lambda=1/2m_{DM}c^2$ where m_{DM} is the hypothesized photon's DM mass. $m_{DM}=(2h\lambda)/c^2 =1.47449943138304 \times 10^{-62}$ kg

Possibly the minimum size of a single DM particle.

Such a prediction will limit the electromagnetic frequency waveband to possible DM group sizes.

1.7. Particle Formation

1.8. Wave Structure Theory

William Clifford in 1870 proposed at the Cambridge Philosophical Society that matter is simply composed of a three dimensional wave nature in Riemann space. Currently know as as Wave Structure of Matter (WSM).[4]

Einstein described matter as spherical and spatially extended and thus the Electron was not a point particle, but rather, a three dimensional structure. Quoting Einstein,

"Physical objects are not in space, but these objects are spatially extended. In this way the concept empty space loses its meaning. Since the theory of general relativity implies the representation of physical reality by a continuous field, the concept of particles or material points cannot play a fundamental part, nor can the concept of motion. The particle can only appear as a limited region in space in which the field strength or the energy density are particularly high.

Since the theory of general relativity implies the representation of physical reality by a *continuous* field, the concept of particles or material points cannot play a fundamental part, nor can the concept of motion. The particle can only appear as a limited region in space in which the field strength or the energy density are particularly high."[5]

Dr. Wolff describes particles as Spherical Waves creating matter. Thus the fundamental particles are spherical wave centers. Dr. Milo Wolff described the electron spherical wave solutions mathematically as:

$$\mathbf{AMPOUT} = (1/\mathbf{r})\,\mathbf{A_o}\mathbf{e}^{(i\omega t - ikr)}$$

and for the positron as:

$$\mathbf{AMPIN} = (1/\mathbf{r})\,\mathbf{A_o}\mathbf{e}^{(i\omega t + ikr)}$$

Where $\mathbf{r}$=radial distance, $\mathbf{t}$=time, $\mathbf{A}$=is the wave's amplitude, ω=angular frequency, $\mathbf{k}$=wave number.

The last equation's $(\omega t \pm kr)$ part, or $\pm kr$ determines if the quanta is an electron or a positron. The sign for the $\mathbf{kr}$ component determines if the waves are traveling inward or outward.[6]

See reference 6 for the complete mathematical modeling.

Dr. Wolff was searching for a physical model matching his mathematical description. The inward traveling waves required a model difficult to conceive. This paper provides a model.

1.9. Wave Structure Model

A WSM model is adopted in this paper.

DM and energy are theorized as the two fundamental "constituents" of all matter. DM is mass with a gravitational field. DM hosts energy in the form of kinetic or potential energy and possess momentum. DM is the source of the gravitational field.

A DM's kinetic energy change described mathematically as either the 3rd or 4th velocity derivative may create gamma rays which are the source of EM fields. The photon is thus associated with the gravitational field.

1.10. Required DM Optical Density

1.10.1. Mini BH Ruled Out

A higher index of refraction than observed in the universe from DM is required for this theory. The first candidate are DM mini Black Holes electron size in mass. Such a mini BH will have a photon sphere, thus assuming all the mass is concentrated at an electron size volume, the radius of the photon sphere (Schwatzfield radius) is calculated:

$$\mathbf{r} = 3\mathbf{Gm_e}/\mathbf{c}^2$$

$$= 2.02934727282288 \times 10^{-57}\ \mathbf{meters}$$

where $\mathbf{r}$=meters, $\mathbf{G}$=Gravitation Constant, $\mathbf{m_e}$=electron mass kg, $\mathbf{c}$=speed of light m/sec

The radius is too small based on electron positron annihilation which yields gamma rays with wavelength 10^{-11} to 10^{-12} meters. Thus mini BHs are not candidates.

1.10.2. DM Phase Change Candidate

A second explanation is a DM phase change forming condensed spheres analogous to water forming spheres. A physical size limit exists, limiting the total possible combinations. These spherical spheres are theorized as transparent with sufficiently high index of refraction enabling photon capture.

The captured photons require sphere sizes that support an interior standing wave. Two photons cannot occupy the same space unless their combined wavelength satisfies the interior standing wave.

Photons are oscillators where the first half of the photon wave's electric vector shares energy with a magnetic field vector. The second half of the photon wave has an opposite direction **E** vector from the first half wavelength. Both **E** vectors do not exist simultaneously. Thus the photon is a totally outgoing **E** field followed by a totally ingoing **E** field as referenced from the center poynting vector.

1.10.3. Origin of Charge

Charge, according to Dr Milo Wolff's thesis is an ingoing or outgoing electric field vector, ingoing for negative charge and outgoing for positive charge.

1.10.4. Lens Effect Reverses E Field Direction

A lens has the effect to reverse the image, thus internal to a ½ wavelength diameter DM sphere the second half of the photon electric field is reversed upon internal reflection resulting in the same direction as the first half of the photon's wavelength. Shown in Fig. 5.

The result is a positively or negatively charged **E** field forming a particle consistent with the electron or positron. The particle posses all the attributes of a DM mass and a photon combined-mass, magnetic field internal momentum, spin etc.

The proposed model requires the photon's wavelength to be at a position wherein all of the photon's oscillating energy is contained in the **M** field when entering and reflecting off of the DM surface. The photon's wavelength must be synced spatially to the DM sphere. This model also requires the DM dimension to be multiples of the gamma's half wavelength. In order to support the DM standing wave, the interior DM dimensions must be multiples of a whole number gamma's half wavelength starting from one.

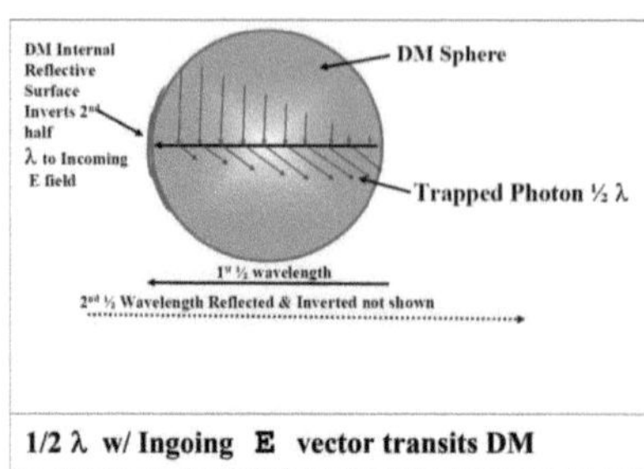

1/2 λ w/ Ingoing E vector transits DM

Lens reverses Outgoing E vector to Ingoing

Figure 5. DM w/ Captured Photon Creating Charge

1.10.5. Electron Positron Discussion

The electron positron is theorized to be formed by a gamma ray trapped inside a half wavelength diameter DM sphere. The entering wave direction determines the charge. If the incoming half of the wavelength has an outgoing electric field vector, the result is a position. If the gamma ray's entering half wavelength has the incoming electric field, the result is an electron. The probability of a positron or electron is 50/50 resulting in an equal ratio of electrons to positrons.

The index of refraction difference between inside the DM sphere and exterior must be high enough to support Total Internal Reflection (TIR).

TIR follows the equation:

$$\theta_c = \arcsin(n_2/n_1)$$

where θ_c is the critical angle between the reflecting surface and vertical to that surface, n_2 is index of refraction inside the DM sphere, and n_1 is the index of refraction exterior to the DM sphere which is expected to have a value of one.

Earlier a DM sphere with electron size mass and gamma ray half wavelength in diameter was calculated to have a density 730 times uranium, indicating the optical density is very high.

Figure 6 shows a gamma ray entering the sphere wherein all of the photon's energy is contained in the M field synchronizing the half wavelength to the sphere's diameter.

Inside the sphere, the outgoing E field collapses the M field until all of the E field energy is contained in the M field. The M field will collapses generating an ingoing E field; however, the lens effect reverses the ingoing E field into an outgoing **E** field. The result is a DM sphere with an outgoing **E** field consistent with a positron.

The process of generating an ingoing electric field and being reversed by the interior reflection of of the DM sphere continues providing a stable DM sphere with an outgoing electric field.

If the gamma photon entered the DM sphere with an ingoing **E** field, the result would be a DM sphere with an ingoing **E** field such as an electron.

If the entering half wavelength is ingoing, the induced second half outgoing field is inverted by the reflection to also be ingoing, and the process continues resulting in a DM sphere with a trapped photon having an exterior incoming electric field.

Note the electron as described is an oscillator based on the photon's frequency.

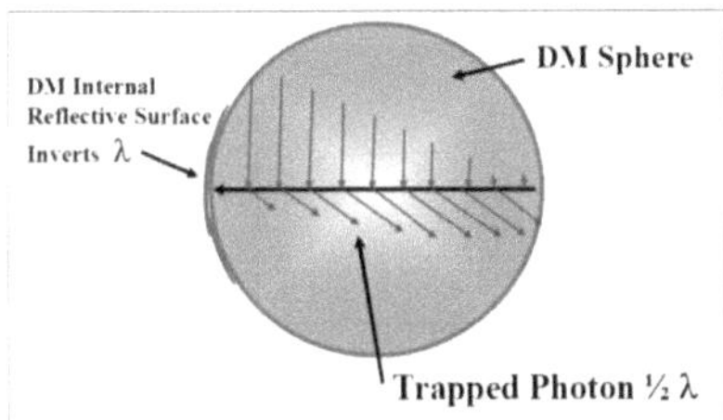

Figure 6. Trapped Photon in ½ Wavelength Dia. DM Sphere

1.11. Possible Particles

Particles must satisfy wavelength considerations as well as possible DM sphere sizes. As the DM sphere size becomes larger more degrees-of-freedom are created for an array of particles each having multiple photons.

Thus particles of one half and one wavelength diameter are the most simple and illustrate the process.

However, larger sphere diameters can accommodate multiple photons that reflect off of the inner DM sphere's circumference similar to a wave guide. The DM sphere's diameter must still support a standing wave which determines DM diameter. Without the inner standing wave the particle does not exist, having an extremely short lifetime.

The particle models described herein have a spherical traveling EM field external to the DM sphere and a spherical standing EM field internal to the sphere, as well as possible lack of external EM traveling field, only an internal EM standing field.

1.11.1. Half λ Diameter DM Combinations

The electron model described above has two possible combinations with one gamma ray (outgoing or ingoing

E field) and one DM half wavelength diameter sphere.

Positron electron annihilation sometimes yields three gamma rays suggesting that our electron may contain two gamma rays. A second gamma ray is possible if it is perpendicular to the first gamma ray. The second gamma ray may be in phase or out of phase with the first gamma ray. A DM with two gamma rays yield three possible combinations, a particle with an outgoing **E** field, an ingoing **E** field, and no exterior **E** field. In all cases, an inner standing wave is maintained. Thus three different potential particles.

The second gamma ray does not increase the particle's charge evidenced by three gamma rays sometimes occurring during positron electron annihilation.

If the two captured gamma rays are out of phase, the particle is changeless because the external field is canceled; however, the inner standing wave is maintained. Thus an electron size chargless particle is predicted.

1.11.2. One λ Diameter DM Sphere

Expanding the DM sphere to one wavelength diameter has the effect of creating a particle with a half outgoing **E** field and a half ingoing **E**

field which cancel the outgoing **E** fields similar to a neutrino. Figure 7.

Such a particle will inter-react very weakly with other particles making them difficult to detect.

The entering photon will maintain opposite pointing EM vectors crossing the sphere with every third ½ wavelength photon with reversed EM vectors. The result is two consecutive half photon wavelengths EM fields pointing in the same direction with the consecutive 3rd and 4th half photon wavelengths pointed in the opposite direction.

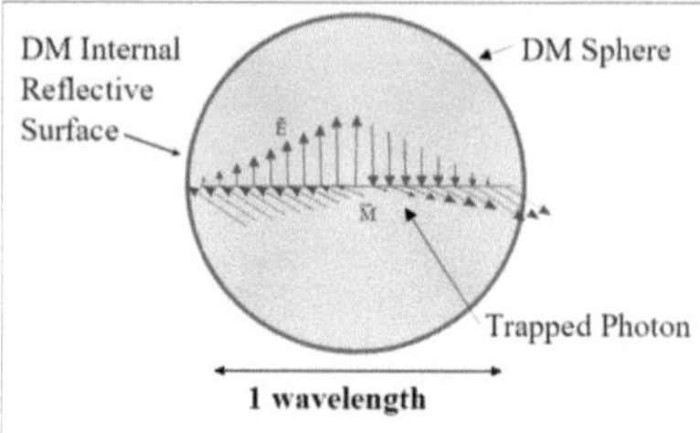

Figure 7. Trapped Photon in One Wave-length Diameter DM Sphere

The external EM fields cancel, but the internal DM structure hosts a spherical standing wave-possible neutrino.

Increasing the DM sphere size allows for additional degrees-of-freedom allowing formation of an array of different particles with multiple photons.

1.11.3. DM Sphere Supporting Several Half Wavelength Internal Reflections

Consider the geometry where a DM sphere can support multiple half wavelength chords reflection off of the inner DM sphere. An example shows multiple, internal half wavelength reflections in Fig. 8. This geometry results in a uniform external field positive or negative based on the ingoing or outgoing phase condition determined by the photon's phase when the photon entered the DM sphere. The second half of the photon's oscillation has its direction reversed via the sphere's lens effect making the **E** field in the same direction as the first half of the wave. This model is also supports multiple photons which may be in or out of phase. Out of phase photons inside the DM sphere cancel the external EM fields while maintaining an internal EM standing wave. The photons can circumferentially reflect internally in the DM sphere and not be restricted to wave number conditions if the DMs circumference is greater than one wavelength.

1.11.4. Figure 8 Detailed Description

The red chord lines on the left show a photon reflecting off of the inner surface of a DM sphere. The DM sphere is four wavelengths in diameter. Internal to the DM sphere exists a standing wave. The **E** field from the reflecting photon has it's **E** field direction changed at each reflection caused by the lens effect which aligns the external **E** field to the original first half wavelength direction, shown as outward in Fig. 8.

The example can host multiple photons allowing for multiple degrees-of-freedom that define multiple particles.

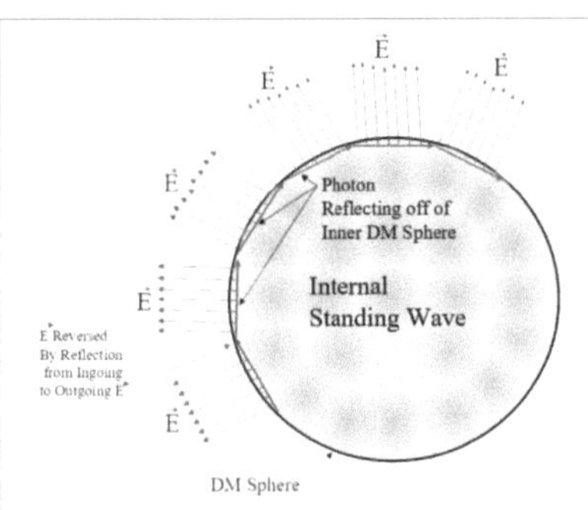

Figure 8. DM Spheres Supporting Multiple Internal Reflections

This model is another candidate for the electron and positron or the proton.

A very large number of combinations become possible with the photon following the inner DM contour via multiple reflections similar to an optical fiber. The possibilities limited by the maximum DM size.

1.11.5. Neutron Decaying to a Proton

The current Standard Model describes the neutron being composed of one up quark, two down quarks, and the proton composed of two up quarks, one down quark. Both particles also contain gluons.

The neutron is also known to decay into a proton and an electron with some missing matter theorized as a neutrino. The quark composition is not supported by the observed decay products.

An alternate theory is that the neutron is composed of a DM sphere containing out of phase photons. The out of phase condition cancels the external **E** field maintaining only the internal standing wave field like a neutron.

If this neutron loses a photon with an ingoing **E** field it becomes an electron type particle, and the remaining DM sphere becoming a proton with its outgoing photon **E** field now dominant. Figure 9.

The neutron is a product of a positive and negative charge, formed by ingoing and outgoing **E** fields, both being in the same DM sphere which creates a neutral charge.

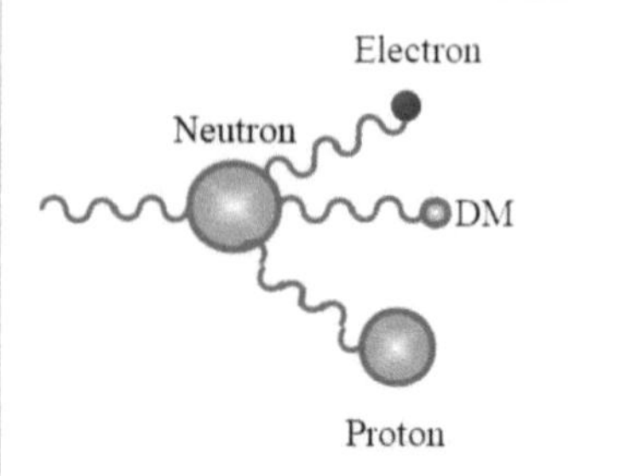

Figure 9. Neutron Decaying into a Proton, Electron and DM

In conclusion, the neutron is postulated to divide into an electron size DM droplet, a proton size DM droplet, and a DM droplet lacking internal photons.

1.11.6. Antiproton

This theory also predicts the neutron can decay into a negative proton and a positron. The antiproton, **p̄**, considered the antiparticle to the proton, fits this prediction.

2. References

1. Tansu Daylan,1 Douglas P. Finkbeiner, Dan Hooper, Tim Linden, Stephen K. N. Portillo, Nicholas L. Rodd, Tracy R. Slatyer. Title: "The Characterization of the Gamma-Ray Signal from the Central Milky Way: A Compelling Case for Annihilating Dark Matter" NASA arXiv:1402.6703v1 [astro-ph.HE] 26 Feb 2014

2. Provided by University of Warsaw, "The birth of quantum holography—making holograms of single light particles," July 19, 2016, Web address: <https://phys.org/news/2016-07-birth-quantum-holographymaking-holograms-particles.html>

3. Libretexts Physics 29.4: Photon Momentum Web address: https://phys.libretexts.org/Bookshelves/College_Physics/Book:_College_Physics_(OpenStax)/29:_Introduction_to_Quantum_Physics/29.04:_Photon_Momentum

4. Clifford, William (1870), 'On the Space Theory of Matter', Cambridge Philosophical Society, 2, pp.157-158.

5. On Truth & Reality, "The Wave Structure of Matter (WSM) in Space" Website: https://www.spaceandmotion.com/wolff-physics-thinking-mind.htm

6. Milo Wolff's Quantum Science Corner's, The Wolff Papers, Website: https://mwolff.tripod.com/point.html

3. Figure of Citations

Figure 1 Author's own work

Figure 2 Author's own work

Figure 3 Author's own work

Figure 4 Author's own work

Figure 5 Author's own work

Figure 6 Author's own work

Figure 7 Author's own work

Figure 8 Author's own work

Figure 9 Author's own work

YOUR KNOWLEDGE HAS VALUE

- We will publish your bachelor's and master's thesis, essays and papers

- Your own eBook and book -
  sold worldwide in all relevant shops

- Earn money with each sale

Upload your text at www.GRIN.com
and publish for free